WATER SUPPLIES
FOR RURAL COMMUNITIES

Water Supplies for Rural Communities

COLIN and MOG BALL

IT Publications 1991

Intermediate Technology Publications
103/105 Southampton Row,
London WC1B 4HH, UK

© VSO 1991

ISBN 1 85339 112 3

Cover photo: Sue Eckstein

Typeset by Inforum Typesetting, Portsmouth
Printed in Great Britain by SRP, Exeter

Contents

Foreword		vii
1	Introduction	1
2	Guidelines for successful water supply projects	6
3	Understanding the process and products	9
	Preparation	9
	Delivery	19
	Additional dimensions	33
	Sustainability	38
4	Four case studies	43
	Malawi	43
	Indonesia	45
	Ghana	48
	Nepal	50
5	Conclusions	53
References		55

Foreword

It is generally acknowledged that improving water supply and sanitation in the world's poorest countries is at least as much about people — their judgement concerning their needs and their participation in meeting those needs — as it is about technology.

So there is value in a report such as this: describing actual people-participation at grass-roots level, and starting to produce a checklist of what is necessary if indeed water and sanitation improvements are to be used and are to be lasting.

What conclusions might those involved in planning and implementing water supply programmes draw from this report? Not, I suggest, that all programmes should demonstrate a highly proficient modus operandi concerning people-participation. That would be unrealistic. Rather what is needed is to ensure that water supply programmes are seriously feeling their way towards such participation; and then to hope that by attaching thoughtful volunteers or other development workers to them, the process will be advanced.

David Collett
Director, Wateraid
July, 1991

Collecting water from an unprotected source prone to contamination, Ghana VSO/Sally Burrows

1
Introduction

There is something inherently appealing about water supply projects in development: water is a basic need, and the results of a project are highly visible and easily measurable. This story, from a VSO volunteer hydro-geologist based in Malawi in the early 1980s, illustrates all the attractions of this type of project:

. . . it was one of the most moving moments in my time overseas – the first borehole that got drilled in the valley. It was quite close to my base camp and as I was on my way to meet some villagers further on, I called in because I knew it was due for testing. And all this muck was coming out, so I knew it would be several hours before the testing and cleaning process was finished.

When I came back the rig was still pumping water out of the ground, this great arm chugging up and down and huge spouts of water coming out, and all the village children were playing in it, throwing it over themselves. This was a village that had to walk five miles to the river; every year was seeing the river retreating because of soil erosion; and the storage upstream was getting less and less. Every dry season they were having to walk further and further.

And suddenly there, right in the middle of the village, was this water! The women were there, collecting buckets, and the children were playing. I can still see in my mind the evening light sparkling on the water. Women who were normally so down-trodden and hard-working encircled me and started to sing. One expression they put into the chorus was 'Thank you, mother'. It was tremendous, you know; brilliant. They were hugging me, and hugging one another and laughing. It was a celebration because clean water had arrived.

I have had similar experiences since and these have brought home to me what clean water means to people. But that was the first time I had been so close to the initial emotion of euphoria in the women when they saw that all this talk and action meant that they actually had water.

They couldn't believe their good fortune. It wasn't thanks to me at all, but I was the symbol on which all their thanks were heaped.

Moments like this will be familiar to people who have worked on water supply projects. They, and the volunteer speaking above, recognize that the reason for jubilation is not so much the possession of the water in itself, but the realization of the potential it offers for improvement in many aspects of the quality of life of those who have it.

Potential

The first, and perhaps most dramatic potential of the possession of clean, accessible water is to improve levels of health of the whole community.

Speaking at the United Nations General Assembly during the launch of the International Drinking Water Supply and Sanitation Decade in 1980, the Director-General of the World Health Organisation (WHO), Dr Halfdan Mahler predicted that 'the number of water taps per 1000 persons will become a better indicator of health than the number of hospital beds'.[1] It is easiest to understand the basis of this potential for health improvement by looking at the threats to health when water supplies are inadequate and dirty.

First there are diseases which come from contaminated water, where infectious agents from human excreta lodge in the soil or are washed into the water source. Faecal-orally transmitted diseases include cholera, typhoid, dysentery, diarrhoea and infectious hepatitis, plus parasites such as roundworm and whipworm. Then there are diseases which result from having insufficient water for personal hygiene — skin diseases such as scabies, yaws and ringworm and eye diseases such as trachoma and conjunctivitis. Diseases spread by insects that breed in water and bite near it include sleeping sickness, malaria, yellow fever and river blindness. Invertebrate organisms which live in water and act as vectors transmit diseases: schistosomiasis and guinea worm are examples. At the beginning of the International Decade, the World Health Organisation estimated that 80 per cent of all disease in the world was due to insufficient or poor water or sanitation.[2]

Many other aspects of community advancement and

INTRODUCTION 3

improvement are implicit in the volunteer's account. Here was plenty of water, in the middle of the village. The time it had previously taken to walk to the water, wait, perhaps in a queue, and carry it back to their homes, was now available for the women to do other things. Those things might themselves be made possible by the water supply: vegetable gardens, with their own implications for health improvement, and income-generating projects like growing crops or raising chickens for sale.

Realizing the potential of a water supply, however, is not an automatic consequence of its arrival. Also writing at the beginning of the International Decade, Professor John Pickford of Loughborough University of Technology noted that it was obviously about water and sanitation, but it is equally about people.[3]

VSO and water supplies

This account draws on the experience of volunteers from Voluntary Service Overseas who have worked in rural water supply projects. VSO has been involved in such projects for more than fifteen years, sending over 40 volunteers to countries worldwide, including major programmes in Nepal, Malawi, the Pacific and Kenya. Such programmes often fit closely with the stated objectives of the organization. VSO aims to increase self-reliance, choice and opportunity and release human abilities in the poorest countries. There is a clear correlation between poverty and lack of access to sufficient clean water, and community involvement in a water project can encourage self-reliance. Writing about the experience of another organization involved in water programmes, Arnold Pacey noted, 'Oxfam has come to see water supply projects, along with others involving house-building, sanitation and other community-based activities, as being potential catalysts for social development.'[4] And the reason that they worked, as he and others remarked, was that the water provided a central, material focus on which community development, aimed at other improvements in the quality of life, could be based.

Many VSO volunteers work in a community development rather than a 'welfare' context. The objectives of the organization state that the specific skills of volunteers working in such a context should support the efforts of local people. These volunteers live and work alongside local people, often targeting their efforts at the poor and disadvantaged or at enhancing the status of women. They aim to share their skills and hand them on to local people. All of this can be achieved through water supply projects.

Community development is easier in theory than in practice, and successful approaches are rarely directly transferable from one cultural setting to another. But by drawing on the practical experience of volunteers who have worked in various settings, using different skills, this account attempts to pinpoint the kind of practices which are likely to result in successful outcomes. It should be borne in mind, however, that the VSO volunteers who have contributed their experiences have spent only two or three years in the projects they describe and they are thus not usually in a position to evaluate whether water supplies have achieved their potential over the long term. In some cases, however, there has been some evidence available from people who have revisited areas where they have worked.

Process

When the process is right, the people who will use the water will understand its potential, have a system which suits their needs, and be able and ready to assist in sustaining it.

It is easiest to understand what the right process is by looking at what happens when it has not been got right. If communities do not or cannot maintain or sustain their water, supplies may become contaminated and delivery systems break down. Thus potential developments may never materialize. In 1980 the Ross Institute noted that 30 per cent of Third World rural water systems were out of order[5] and an International Water and Sanitation Centre report estimated that between 40 and 80 per cent of handpumps became inoperative within three years of their installation.[6] Yet if the potential benefits of the supply are to be realized, the water has to be continuously available.

Throughout the International Decade, two elements were seen to be central to the successful process. One was *the involvement of local people* in the design, construction and continuing care of their water supply. Only through community involvement and active participation can a situation be created where the water and its supply system are seen as belonging to the community (rather than being artificially created for it or thrust upon it by external agencies). Desirable though this is, it is not easily achieved: Lindsay Robertson, community development and water supplies expert, long involved in a highly successful community-based programme in Malawi, believes that it takes ten years of hard work to achieve successful 'community ownership' of a water supply.[7]

The second key was *the use of appropriate technology* in the design of water supply systems, so that local people or maintenance teams could service and repair faults without the need for complicated and expensive imported parts. Again, however, this 'ideal' is not easily achieved, nor even possible in some circumstances.

Although there have been some reservations expressed about the effectiveness of these key features as a guarantee of continued supplies,[8] they remain central ingredients of success in delivering clean water to everyone who needs it.

2
Guidelines for successful water supply projects

Drawing both on VSO's experience (as described in the following two chapters, and that of other development organizations involved in water supply projects (as reported in the sources listed in the References), the two observations with which the previous chapter concluded can be extended into a longer list of 'guidelines for success' in water supply projects. In this context success means projects which lead to the provision of water to rural communities which is of sufficient quality and quantity to lead to sustainable improvements in individual health and in other related aspects of the quality of life.

GENERAL CONTEXT

o Programmes for water supplies, and local projects within them, should be sensitive to local cultural and social traditions. They should be mounted in response to genuine popular demand.

o Projects should, wherever possible, take place in the context of health promotion activities.

PREPARATION

o The active participation of the community in the process which will lead to the design, installation and subsequent maintenance of the water supply system should begin at the earliest possible stage.

o Initial approaches to rural communities should be made through traditional and established leadership. A village or community water committee should then be established, by or with the local leadership. This should be linked to or should build upon existing local structures, and should involve both men and women, if possible. Alternatively or additionally, a separate, or parallel, women's committee may well be appropriate.

o Adequate time should be allowed for preliminary consulta-

GUIDELINES FOR SUCCESSFUL PROJECTS

tion and agreement with the water committee. In some circumstances, a demonstration project may be appropriate, so that the communities get some idea of what the practical possibilities and limitations will be.

o The local community should be involved in the design of the system, so that local traditions, knowledge, skills, convenience, and technical possibilities can be taken into account.

o Preliminary health education is essential. It should include the identification of local volunteers to work on campaigns aimed at specific groups in the village, such as children, mothers and old people.

o The community should be clearly informed, through the water committee, of the demands which will be made on it and responsibilities which will rest with it, before, during and after construction.

DELIVERY

o Health education activities should continue through the delivery component of projects.

o The local community should be involved in the construction of the system as much as possible. Contributions of time, labour, food or money may be sought as possible and as appropriate, but should be geared to the community's ability to contribute.

o Where projects form part of larger water programmes, particular members of programme staff should be associated with particular projects. Wherever possible this association should continue through all three phases — preparation, construction and maintenance — of the project.

o The system installed should use the simplest appropriate technology so that it is capable of being maintained and repaired by local people. Wherever possible, the system installed should use locally manufactured or available equipment.

ADDITIONAL DIMENSIONS

o The local community should be encouraged to explore possible uses for the new water supply, for example in agriculture, and in economic development activities.

o The committees established to prepare for, help construct

and thereafter maintain the water supply may be appropriate vehicles for other development activities capable of being pursued by communities.

SUSTAINING THE FLOW

o Health education activities should continue after the installation of the water system.

o Village or community water committees should have an agreed plan for collecting money to support the maintenance of the water system.

o Communities should appoint official 'caretakers' for the new water system, some of whom should be women. These caretakers should have access to continual training, support and advice.

o Parts for water systems should be readily available, and tools for repair should be in the custodianship of the water committees.

o Water supply projects should be evaluated. Appropriate measures for such evaluation could include the extent of upkeep of the system, and consumer response, as well as maintenance costs and health indicators.

3
Understanding the process and the products

The experience of VSO volunteers working in water supply projects in rural areas can best be described and analysed by seeing the projects as consisting of four components: preparation, delivery, additional dimensions and sustainability. Each is now discussed.

Preparation

Recognizing the need

'If the initiative does not come from the community itself, but from other sources, the project usually meets with failure after construction is complete.'[9]

While this is easier said in theory than achieved in practice, it introduces and highlights one aspect of water supply projects that is of paramount importance: the participation of the communities in their projects. If the people who comprise the community are to derive the maximum benefit from the project, they must feel that it is *their* water, that they have played a role in bringing it about: that they *own* it, in other words. It has to be remembered that many communities have an existing water supply of some sort, even if it is unsatisfactory. Its limitations, or even its dangers, may not be apparent to many users: water which causes ill health may not *look* any different to clean water. A new water supply will often require some sort of change in daily habits, and communities may often not feel comfortable about the idea of a change in their way of life. The greater their sense of ownership of the new supply, the less this particular problem will be, and the earlier this sense begins to be engendered, the better.

What this points to is the importance of work with rural communities to raise consciousness about the various potentials of an improvement in water supply, well before any consideration of the development of new supplies is made. The major differences in experience between the projects described by VSO volunteers lie in the existence or not of this preliminary, pre-project phase, and the length of the time allowed for it. But a dilemma lies in deciding whether work with communities to help them in the recognition and exploration of their needs should be of a general nature, or in the stricter context of water supply. A volunteer who had worked on a government programme in Uganda which had a limited preliminary phase, said, 'When we would go to interview communities about having a spring protected, they would say, "That's all very well, but we want other things." Water was very low on their list of priorities, and we were arriving and telling them that they needed it. Everybody in Uganda has water. It's not clean, but they have been using it for centuries. People will say, "My old dad used it, and he never got sick." How can you argue and say, "We want you to spend time and money on protecting your water supply"? They will say, "What for? I would rather spend the money in having iron sheets on my roof so I don't get wet at night." People must perceive the need for themselves.'

In some areas the need for improved water had emerged from community development projects, where a community worker (sometimes a volunteer, sometimes a local person) had been working for some time with a village, helping people to examine all their needs and exploring how to meet them. In these circumstances, of course, there is no guarantee that water will emerge as a priority, but the indications are that, if this kind of work is combined with a health education programme, it is likely to.

But such broad community development processes take a long time, and this is not necessarily available to programmes which are funded with the express purpose of improving water and/or sanitation, often by donors who are anxious to see targets for supplies met. Nevertheless, programmes of this type often can and do recognize the importance of community need-recognition, leading to expressed demand, by working *only* in villages which ask for supplies. But even so it is important to recognize that there is a difference between a *request* from a village for new water, and a widespread *perception* of the need for that water in the community.

In Nepal, for example, where a government programme responds to requests from traditional village councils, volunteers noted that requests could be made without any consultation among or participation by, the whole community. They also pointed out that requests were not always based on need: 'The politics of villages acquiring a water system were long and complex, and nothing to do with us, as technical volunteers. But it seemed that the prestige value of having a water supply project could override the issue of whether one was really needed.'

When this happened, the volunteers found that it was difficult to complete the water system, because the people who felt they had not been consulted about the supplies, were nevertheless required to contribute to them. Thus the absence of early community participation in need-recognition can hinder subsequent participation in delivery.

Worse, if those who need the water are bypassed, and others, with less need, are supplied, water supply can prove divisive. A volunteer remembers: 'We had been told of the importance of the community role in water supplies many times in theory, before we went to Nepal. If the influential part of the village managed to get all the pipelines running to it, with the communal tapstands in that area, then there was a good chance that the pipeline would be sabotaged. Until you have come across this at first hand, it is hard to appreciate that it can happen. It is extremely important to understand and sort out these community dynamics before you start installing a system.'

Another volunteer, in Indonesia, came across similar problems: 'We sometimes went into a village before the questions which needed to be asked had been asked. Whose idea was the scheme? Why does he want it? Ideally you want initiatives to come from the grassroots, and then you support these initiatives, but on the other hand, you are anxious for the right initiative to come along, you are trying to nudge it in the direction you want it to go. There is a conflict in that.' In this instance the 'grassroots' enthusiasm was for a water supply scheme (check-dams to hold water from the rainy season through the dry season) which turned out to be technically inappropriate for local circumstances: 'You had to understand why there were so many enthusiastic requests. One reason was that to have something as big as a dam in the village was prestigious — powerful people would come to look at it, and congratulate the village leaders for obtaining it.'

Thus while the 'ideal' approach involves broad community participation in identifying the need for water and what form the supply might best take, at the same time involving the appropriate external agencies from the earliest possible moment, this is not necessarily achievable. Often, for example, externally planned programmes may only begin to discuss needs and ideas with the people after their request has been received. This was the pattern for a non-government organization working in another area of Indonesia: 'First would come the request from the village. Then there would be negotiations with the people as to what they could provide for themselves. These were carried out by field staff and, depending on what was available in the village, and what the motivation was, a judgement could be made as to which villages would have supply projects in the next financial year. The ideal would be for the organization to provide the technical input, and the villagers everything else. The ethos of the organization was to work with the poorest of the poor, but it was recognized that in many villages the poorest people are not at the right stage to be able to contribute to a scheme. To work in a village because the people were the poorest would not necessarily be successful.'

Building on what people know, and on existing community structures

So what helps secure community participation? Two factors emerge from VSO's experience. First, building upon what people, and the community as a whole, already know: 'Their ideas and their attachment to old ways must be incorporated into projects. Above all, what is good in their present water and sanitation practice should be accepted and developed, rather than being replaced by the latest designs from some international organization'.[10]

Many VSO volunteers have recognized and valued the existing knowledge invested in the communities where they have worked. A volunteer in the Solomon Islands, for example, said: 'People usually have a good idea of which source they want to develop. They are usually right — they know the ground. They know the nearest source of good water, and they usually know what is good water.'

An important aspect of this concerns land use and land rights. Failure to recognize this aspect of local knowledge can cause problems for the design of water supplies, and in consequence some programmes concentrate on improving existing sources, rather than finding new ones: 'We started off with existing wells – the user group is immediately identifiable. Also, you can be sure, by talking to people, whether there is water there or not. You can build on the people's expertise.'

Secondly, it is important to discover and make use of the community 'government' structures which exist, for these will very likely have participative dimensions, and can thus be useful ways of consulting the community and its leaders. Whatever committee or other structures are then established to oversee the conduct of the supply project may then be built on or around the existing structures. In Malawi a volunteer reported that 'the first visit was always to the village headman'. Trusted leadership may be very helpful in mobilizing villagers to support the construction phase of a project, but health education and recognition of the potential of a supply will also need to include the village leadership. The key is to understand the existing leadership and 'government' structures, so as to recognize their strengths and limitations. For example, in some cultures a great drawback lies in the fact that leadership and other structures may exclude women.

Involving the whole community

'Water is women's business.'[11]

Although the first discussions about water supply may have to be with male community leaders, men often know very little about water. While by the end of the International Decade this was well known, it was not widely recognized at the beginning. This experience comes from a woman volunteer in Nepal: 'We would announce ourselves to the village headman and ask basic questions. "Where do you get your water? Can you show us?" and they would take you to see and . . . the water hole had dried up. They hadn't appreciated that it was the dry season, that that hole always dried up in the dry season, and that their wives used another, miles away, at that time of the year. Or you would get to the hole and there would be women there. You asked the women what they used the water for, and they said, "We don't drink this, this is too salty — we only use it for washing." Again, the men did not know. I learned early on that if you really want to know about water usage patterns you ask women, not men. Village headmen are the first point of contact. You have to go through the proper channels, but if you want to find out what the score is, you have to talk to women.'

In a programme in Ghana which led to the installation of clean water supplies, the initial focus was entirely on women and their needs. It was started by a woman volunteer community worker. On the advice of a local woman teacher, she visited the village elders to explain that the need for improved water had emerged from her discussions with the women: 'They were a bit suspicious, but I explained the women's health needs and they agreed to allow the women to organize a committee, within agreed guidelines. Some families were closer to the chief, and it was important that they set the lead and were included in the planning. There are hierarchies in the villages and they must be respected.' Thus, although it is essential for any water supply project to acknowledge the position of women, special efforts often have to be made to include them in the planning of projects because of their traditionally subordinate position. They bear the responsibility for the quality and quantity of water used in the home, for household hygiene and refuse disposal. The responsibility is time-consuming, as we have seen, and it is also physically exhausting. Research has shown that women in Third World countries expend 9 per cent of their calorific intake on carrying water, and that rises to 27 per cent in mountainous areas.[12] The physical burden of carrying water may damage a woman's spine and pelvis and can lead to complications in childbirth.

So it is vital for women to be consulted about and involved in water supply projects, however difficult this may be. In seeking and securing their involvement, however, clarity and realism about appropriate roles for them are essential. In negotiating the village contribution to the construction phase of a project, it may seem counter-productive to include women in the process. In Nepal, for example, 'Women have no status. It would be nice to have a woman involved on a water planning committee, but they need to have some status in the village in order to mobilize people to work. To mobilize people you have got to have loud-mouthed, strong-willed personalities who can pull everybody else with them – motivators. Not women.' But there are many cultural variations. In several descriptions of African projects women did appear to be acting as community leaders and mobilizers, for example. And it has to be remembered that preparatory and preliminary consultation with communities should not just focus on the next, the delivery, phase but on longer-term needs and issues, concerned with maintenance and health, for example. In these matters, the role of women will be important.

Employing female staff on water supply projects may be helpful insofar as securing the involvement of local women is concerned. In the programmes where VSO volunteers have been women (whether hydro-geologists, hydrologists, civil engineers, community workers, social educators or sanitation engineers) there has been a higher level of involvement by women, though not always in the preliminary stages. In most cases the volunteers felt that their gender had helped in communication.

Other possible activities which appear to encourage the involvement of women in the earliest phases of a project are preparatory meetings for women only; a number of guaranteed places for women on village water committees; parallel women's committees (since women may be intimidated in gatherings of men); and promotion groups using crafts and skills which women enjoy, and in which they have expertise.

Preliminary health education
Whilst all the foregoing observations indicate that cultural, institutional and other variations mean that the preparatory phase can take a variety of forms, VSO's experience points strongly to the fact that a highly effective focus for preparing for water supplies is a programme of health education undertaken before the supply is planned or designed. This is a way of alerting the community to the invisible dangers of water, and of beginning to introduce ideas of hygiene and habit change which will help to ensure the maintenance of the supply in the long term. In several cases, volunteers mentioned that they felt the water supply would be less effective than it should be because it had been introduced without parallel improvements in sanitation. Some felt that the best operational sequence would move from health education to sanitation to water supplies.

While there was a health education component of water supply programmes in several countries where VSO volunteers have worked, it was usually introduced at the same time as construction, or afterwards. But in some programmes where this had been the practice, a recognition that pre-construction health education was necessary had led to changes, so that

Discussing water needs with the women's group, northern Ghana VSO/Sally Burrows

health educators were 'first in' to rural communities. In Indonesia, a volunteer public health officer was attached to a basic supply programme, and over three years trained staff in aspects of health education and sanitation to go with the water supply.

In the Solomon Islands, where the responsibility for water lies with the Ministry of Health, and where health inspectors are responsible for the initial assessment of village requests, the link between health and water supplies is as a result strong. 'In general the people have a very high level of health awareness,' says a volunteer water engineer, formerly employed in this programme. 'There are lots of health programmes on the radio, and everybody listens to the radio. From the health point of view, quantity is important. If people wash often they are less likely to spread disease. But you don't have to persuade Solomon Islanders to do that: they love water.'

Perhaps the strongest argument for the introduction of health education at the earliest stage is the difficulty of the task it has to effect: changing people's habits. The same Solomons volunteer comments that 'the whole health thing can be very difficult to get across, especially when you are talking about bugs which can't be seen. I heard a story about a project where ferro-cement tanks were being installed to catch rainwater. The engineer re-visited a village where the tanks had been in place for a year or so and found that the water coming out of one of the tanks was brown. It transpired that the people didn't like the taste of water from the tank. They preferred the taste from the river, where they had collected it before, and they had added some river mud to the tank to adjust the taste.'

This is a good illustration of the point made earlier, about the need to find out what is important to the people who will use the supplies, for custom, tradition and usage are important. But it also illustrates how important it is for the introduction of water to take place in the context of health education activities.

Water committees

The central question which emerges from this discussion is: what kind of institutional framework, established at the local level, is needed as the focus for a water supply project? It will need, as has been noted, to maximize community involvement and participation, to ensure that local knowledge and existing structures are built upon, and to ensure that health and other benefits accrue to the community.

What emerges strongly from VSO's experience is that if, at an early stage in a water supply project, a group or committee of local people is established as the institutional framework through which the planning, construction and maintenance of the water system are undertaken, then these goals are more likely to be attained. This emerges from all the countries in which VSO volunteers have worked, although cultural variations inevitably mean that there is no standard 'model'. In Indonesia, for example: 'As soon as we arrived the village would establish a committee with chairman, vice-chairman and secretary. It was usually obvious who the chairman would be, and a successful group seemed to depend on locating the right person. They might be traditional leaders, or they might be people like teachers or priests. They are likely to be better educated yet able to communicate with people.'

In the Ghanaian village where the approach had focused on the needs of women, groups of ten (family) 'compounds' which would use each source formed their own small committees: 'In some places this was very quick, in others it took six months. It depended upon a couple of people having good skills and knowing how to mobilize others.'

In Kenya, where a female water engineer was involved in a programme which led to the development of 21 dams, 22 shallow wells, 5 rock catchments and 267 water tanks in two years, local committees were already well established before she started, and were often chaired by women. In this case, the catalyst for their development had been the development committee of a local diocese.

In Nepal, on the other hand, volunteers had to set up water committees themselves: 'The construction of your water committee is crucial. My idea of the perfect committee would be to have someone from each ward in the village. It is difficult, though, because by the time you have identified who would be suitable, it is too late.'

Committees are therefore best formed, and its members identified, by a local person or agency, rather than a volunteer, for an obvious impediment is that the volunteer's language skills will very likely not be up to the task. So: 'You need a close relationship with somebody who can act as a conduit, who can speak the language and understand the people.'

Many volunteers noted the importance of 'natural' com-

Drilling a shallow tube well, Nigeria VSO/Jeremy Hartley

munity leaders in the setting up of water committees and in maintaining the interest and involvement of the community. Many found that these people were women. Regardless of who, and of what sex, they turn out to be, it is important to find them.

A further lesson that emerges from descriptions of water committees is that the smaller the number represented, the more likely it is that the committee will be supported and effective. In the Solomon Islands a volunteer recalls: 'It was the smaller villages with just one chief which organized themselves best, rather than the larger ones, of 1000 people or so, where it is more difficult to raise community support. There is less squabbling, less population change in the smaller villages.'

Delivery

Maintaining enthusiasm

For many volunteers, the initial contact with a community comes with the need to survey it, find out what the current water situation is, and determine what possibilities there are for improvement.

In some instances, the general nature of the answer to existing problems may already have been decided – it will be a borehole, a gravity-flow system, a tubewell, spring protection or whatever. In others, there may be some scope for flexibility. Where there is a technically qualified volunteer available who can adapt plans to different situations and needs, the result can be cost effective, in that cheaper solutions, and/or alternatives to pre-determined plans may be possible. Certainly it appears that too rigid an approach to the type of supply to be used can be wasteful: 'In the past an overseas NGO provided drilling rigs and staff, drilled holes, put in pumps and went away again. Two years later 50 per cent of these were broken and the other 50 per cent were silted up and giving half the yield they were supposed to.'

That occurred in part of Nepal. In Malawi, a volunteer arrived to work on a similar 'blanket' borehole approach: 'It wasn't very effective, because there had been no technical assessment of whether a borehole was the best approach. Perhaps a hand-dug well would work better, or maybe there was potential for a piped gravity scheme in the area, instead of a groundwater source. That is very fascinating, thinking of solutions to technical problems, and ways to improve design.'

But it demands time and flexibility, which are not always available when programmes are under pressure to deliver results.

In communities where there has been previous intervention to provide a water supply, and the project has not been completed, or has failed soon after completion, it is understandable that there will be some local scepticism about yet another water supply project. Several volunteers described experiences of conducting surveys and designing systems in areas where surveys had been conducted in the past, and nothing had resulted. Gaining and sustaining community interest in these circumstances was understandably difficult. In general, the shorter the period between survey/design and construction of the system, the less the difficulties appear to be. Once expectations have been raised, people look for something to happen.

But it is not always easy to do this, whatever the circumstances. In Nepal, for example, speed of action is limited by seasonal factors. 'Feasibility surveys happen one year, at the driest time, and then, if it looks as though the job can be done, you go along in the following dry season to take detailed measurements, work out a ground profile, look for locations where you can site reservoir points, break-pressure tanks, tapstands.'

If a village water committee is in place, and is encouraging interest in the scheme, it will give the designer a mechanism for consulting with users about the design *details*: such as where people want the tapstands to be, how they will be using them, and what fits best into their routines.

In another project, again in Nepal, training as health educators was given to selected women from a region. 'The idea was that they would go into the parish from the point when we started work there, and stay for the whole time. They would take part in the surveying and site selections, have discussions with the villagers, and remain during and after the construction stage of the project doing health education in schools, and house to house with mothers.'

This kind of approach (whether oriented to health, or to general community development) is another way to sustain interest over the long term, as well as a means of ensuring that the design fits the people's wishes.

Another approach which may overcome scepticism and develop or maintain enthusiasm is the 'demonstration project'. Here, instead of approaching a programme as a series of stages on several sites – survey, approval, ordering materials, construction, and so on – single, small supply systems are designed and built in order to illustrate to the rest of the community what is possible. Another Nepal volunteer, for example, found that there was a hold-up in the middle of one project because an insufficient quantity of pipes had arrived before the rains. 'The morale of the people was low and they did not believe the project would be completed at all . . . I decided to build a smaller system, that had only three tapstands and one reservoir tank and one intake, because rapid and good quality construction could be achieved and this would serve as an inspiration to the rest of the village that this water supplies programme could deliver . . . Water was running at this small project before the monsoon started, and that gave us all confidence that it was actually possible to construct water systems!'

So, work in one part of a rural community acted as evidence of what was possible for the rest of the community. A strength of rural areas is that, despite poor road communications, there is often a 'grapevine' which results in information about what has been done in one village getting spread further afield. As a Uganda volunteer recalls: 'In one particular place, which is very hilly, and has a lot of springs, we got one community to protect the spring, and some people across the hill, who also had a spring, came to help. And they went home and told people, who said "Let's do it". So they came to us and said, "We want one too, when can you come to do it?" '

As far as motivation to *care* for the installed water supply is concerned, it appears reasonable to assume that it will be high in a community which has asked for one knowing, thanks to a demonstration project or through other means, exactly what the result will be. Developing a complete supply system on these organic principles of demonstration, demand and consequent care is not, however, easy, and it will very likely be more expensive and apparently less efficient to construct than a large-scale, planned multi-site programme. But what the demonstration project does offer is a reference point for planners and consumers so that the pitfalls, practicalities and advantages of the system can be grasped. It also enables users to communicate with other potential users – always a good marketing strategy – and, while more expensive, it may nonetheless be, in the long term, more efficient.

Continuing to build on local knowledge

If there is flexibility in a programme, community advice can be responded to, and community knowledge taken into account in the design of the supply. Technical development can increase the scope for flexibility. In one Malawi programme, a standard design had been updated: 'One thing about the new technical design was that you could drill boreholes in a lot of places you wouldn't have dared to put them otherwise. It didn't matter that you weren't going to get much water, because you were going to get eight or nine times as much with each metre you drilled, with this more efficient system. That meant that the community could get involved a lot more with the actual siting of the boreholes. People could decide where they wanted the hole, provided they had some guidelines about places which could not be used.'

Taking local preferences into account can often be an important success factor: 'There had been a survey done for this village. The water source was 150 metres up in the hills, but it was a very small stream, so the water had to be stored in a tank first. The original design meant building a spring box, feeding the water into a tank up there, then gravity-feeding it down to the village. But that meant the villagers would have to cart all the materials up to the top, and they asked me if there wasn't another way of doing it. The village was terraced, with houses on different levels, and gardens at the top. That turned out to be perfect place for a small tank. There was all the pressure needed, small pipes were involved, and it used a small-scale tank which we invented. There was a great deal of interest in this whole process in the village – people got all their relatives to come along.'

Thus although this Solomons project was part of a wider programme, by working in consultation with villagers, it became a unique response, something designed specifically for local needs.

Where design is inflexible, not only is this opportunity missed, but there can be major problems in the system which simply get replicated: 'The Department of Health had produced a booklet full of basic designs, and almost every stand-pipe in the country had been built to that design. But the concrete base had no drainage, and soon got undermined by the water running off.'

Thus no matter whether a flexible or pre-determined, 'custom-built' approach is taken, the technician, whether a volunteer or local field officer, needs to listen to local people and not impose solutions upon them. It can be difficult for qualified professionals to take this role, as a servant of the community, but in the long run it offers a greater likelihood that the villagers will understand and maintain the system.

It should be noted, however, that local knowledge and traditions can work against the plans for a supply, for instance this Solomons example: 'It can be very frustrating at times. The field technicians I was working with went to look at a hill where there were springs coming out, and running into a stream, and then off into the sea. All that year, all the year round, and nobody was using it. But it was on the slope of a very sacred hill, which is why nobody would touch it. Of course, you cannot ignore traditions like that, but it is very tempting at times.'

The hardware

If flexibility is one key to good design, then simplicity is another, especially insofar as the materials and machinery used in the supply system are concerned. As a general principle, the more appropriate they are to local conditions, and the more easily understood they are by local users, the more chance there is that they will be 'owned' and looked after.

This has long been recognized by those who run and develop water supply programmes. Well before the International Decade began it was observed, however, that though technologies have often been too complex for the context into which they must fit, this is probably less of a problem in water supplies than in other development programmes: 'The level of technology used in rural water supply schemes is inevitably rather simple. The idea that the technology is too complex for the simple rural folk of developing countries is a myth and a rather unconvincing myth at that. Broken and inoperative water supply schemes are to be seen in environments where village skills extend to maintaining and repairing bicycles, transistor sets, irrigation pumps, ceiling fans, air conditioners and a variety of small industrial machines and tools.'[13]

But while complex equipment may well, therefore, be maintainable, it often has a high likelihood of breakdown, and parts for it may be difficult to obtain. Even where the equipment is simple, a variety of factors (including those described earlier, about participation in planning and design) may conspire together to produce failure. For example, a volunteer in the Gambia who was not working in water supply, but had made observations about the importance of water in the village where he lived, noted on his return: 'It was very interesting to find that during the Water Decade a German-funded programme had installed a hand-pump way outside the village which was not working very well, and people were still using the traditional water points. The seals had broken on the hand-pump and it was very inefficient, needing a lot of hard work to pump the water. No spare parts had been left.'

In this instance the villagers had not been motivated to repair the pump, there was a fault in the design of the whole system, which was also placed inconveniently, and other factors, such as the absence of a caretaker, and the existence of other sources of water, all contributed to failure.

The results of equipment breakdown vary. If a hand-pump breaks down, no water comes out. If a gravity-feed water system breaks down, water nonetheless runs constantly, but this may be undesirable for health and hygiene reasons. It is water, at least, and people will often accept a less than perfect system. One volunteer in Uganda noted that villagers saw the advantages of handpumps in delivering clean water: 'People like the pumps. They see the water coming out as clean. But this preference is to do with convenience first and foremost. If it is going to be a pain to get the pump fixed, and you can get water elsewhere, people won't bother too much.'

In a survey of villages in the Solomon Islands (conducted for an MSc. thesis) one volunteer asked villagers why hand-pumps had not been repaired. Two reasons were given: a lack of tools and a lack of know-how. But, as the volunteer pointed out, many of the villagers were able to maintain far more complex trucks. Nevertheless, the two clear needs in this example were for readily available parts and for basic training in pump repair to allow village-level operation and maintenance, and these are now being built into the programme.

More difficult to solve is the problem where an essential piece of equipment, sometimes needed for a project in a difficult area, fails to operate, through a design fault. One volunteer working on a design for water supplies from the mainland to a school on an island in the Solomons with no water, incorporated a hydraulic ram pump which he had seen in use in other parts of the country. 'The pump was available, and though it was imported it had been designed for use in rural areas. It is appropriate technology, you need very few tools, and if it is built properly it should work continuously. But the valve components are made of material that is not strong enough and they wear out very quickly.'

In this case the system was redesigned, omitting the pump. Although it supplied water, the amount was less than demand, and as a result the system failed to meet the full need.

In other settings, appropriate technology can work very well when attuned to local needs. In order to cut down dynamic loading on supply systems in Nepal, no taps were used on the outlets: 'That is a good system from a water point of view – there is nothing to go wrong because there are no moving parts.' And at the same time it responded well to the preference of the people for flowing water. They were used to a flowing supply and believed it to be healthier than water that was enclosed.

But however strong, simple and appropriate the hardware in a water supply system is, the pressures on it from constant use by many people are likely to lead to a need for repair. Piped supplies in gravity systems need to be buried, if possible, to avoid damage from animals, falling trees, and people. Several Solomon Islands volunteers learned about the real appropriateness of technology from villagers. 'On some pipelines, where breakages had occurred through land slippage, or vandalism, adequate repairs had been made with bamboo, and the system still worked, although the bamboo gets a growth inside it. I don't know if that affects the quality of the water, but it doesn't affect the flow.'

Using a vertical iron pipe as a drilling implement, and with a helper to lift the pipe up and down by a bamboo lever, a skilled local person uses his hands to provide the suction which bores a tubewell. This traditional method, used in northern India and the Ganges plain, was originally developed with bamboo. The availability of iron pipe and plastic PVC lining pipe means that the traditional method has been improved, and wells can be deeper and be kept cleaner – but the essence of this technology is that it has evolved from a local skill, rather than replaced it.

Training and exchanging

Thus a successful project is not just one which works, but which continues to do so. The establishment of some form of training arrangement, both for the field staff of water supply programmes, and for village users, is important if long-term and proper maintenance and capacity to repair are to be achieved. The use of workshops, to which local people come from a surrounding area to learn particular skills, is well tried in development programmes, though their feasibility depends on the existence of good transport and communication systems. Health workshops are now widespread, and schemes like the UNICEF programme to train local volunteers in the skills of assessing, planning and implementing rehabilitation for disabled people have also demonstrated the general usefulness of the technique. In water supply one Indonesia volunteer described his experience: 'The workshop for water maintenance needs to last for two or three days. It is important that it is completely practical. We get each group to install a new pump from scratch. The main problem is not teaching people these technical skills, it is getting them to learn in a practical way – they are used to a school system which is very formal, sitting in desks. You cannot teach practical skills in that way.'

The other method by which skills in design and construction have been passed on is through the counterpart/apprentice relationship between volunteers and local field staff in the programmes where they work. Often the volunteer has been responsible for training field staff, and in these circumstances both on-the-job and classroom techniques are used.

Participation

'Programmes which involve the widest possible participation of the people whose needs are addressed are most likely to be effective.'[14]

The necessity of this participation at the construction phase is a linking feature of all the programmes in which VSO volunteers have been involved. Although this is closely linked to the longer-term goal of continuous maintenance, there is often another, more immediate, pragmatic reason why it is necessary: 'In the poorer countries, there may be no hope of meeting even modest targets for the provision of services unless there is community participation.'[15]

So getting the community to contribute to the construction of the supply is not just *desirable*. Often it will be *necessary*. If a well organized preliminary phase has been conducted, discussion of what this contribution will be will already have taken place, and agreements been reached. One question which arises concerns payment for community labour. What local people can be expected to do for free, and what they will be paid for, varies with the situation: 'We would pay the villagers for portering and other semi-skilled labour, like mixing cement. Once the village had applied for a water supply, contracts were drawn up between it and the ministry. Villages did not have to pay for the pipes and cement, but they had to provide labour for certain jobs. Some jobs would be paid for: the carrying of pipe from the road head, which could take four or five days, would be paid for, but the actual digging of trenches to bury pipe, or the collecting of stones: that was the villagers' contribution and was not paid for.'

That this labour would be supplied was a formal, contractual condition on which this project in Nepal depended.

But in one area in the Solomon Islands, where the idea of such a formal contract with villages had been recommended by outside consultants, the volunteer felt: 'I can't see it happening, because that is not the way things are done there, and the way they are done works quite well, I think. The village was expected to work with the health staff who were constructing the project, and they knew that would have to be done. Villagers would provide all the sand and gravel. It didn't cost them anything, but it was a lot of work. Some villages are enthusiastic about this, and others are not. The difference seemed to be in the amount of money the villagers were earning elsewhere. If the people are busy working on plantations, they are less likely to help on the water supply. But even the villages which were less well organized still knew how to work together collectively.'

In another part of the Solomon Islands, financial contributions were asked of each community: 'Once we had turned up to survey the village, we would sit down with the local village council and explain the costs of the supply system and the condition for installing it: that the village provide the labour and a cash contribution of 25 per cent of the cost. The people understood this and they always paid. The idea was that by finding that small contribution there would be more onus on the people to look after the water supply.'

In Indonesia, an organization working on water supplies as part of an integrated programme had also learned from experience that a community contribution was not just desirable, but necessary: 'The main idea behind community input to a water system, whether they provide the sand and gravel or whether they contribute towards the cost of cement and pipes, is not that you save money and can afford to do extra projects – though you can – but that people, as they are building it, learn how to repair it. In the first sanitation project in a village we provided everything: the cement, bowls, pipes, the lot. Everybody built one, but few of them were used. That changed the thinking. The (next) project expected greater commitment and input, and though it was not an agreed condition, it was normal that the community would provide latrines, as well as materials and the organizational structure to maintain the system.'

In a women's project, where water had emerged as a concern, the digging of wells was entirely carried out by volunteer labour. Ten to fifteen people worked for ten days on the first well, with men doing the physical digging and women feeding them and carrying sand and stones.

In another project in Nepal, a volunteer found that a hand-dug well programme was being carried out by a paid team. 'At first I felt that was a pity – that the more contribution the villagers could make, the better. But the counter-argument to that was that professional diggers are efficient and quick, and they work safely. If you use a local group, it is not feasible to train them in safe techniques just to dig one well. The wells might go down 50 feet, through clay with stones and boulders. If something fell on someone's head, or the walls came in, that would be a major disaster, for the village and for the project.'

What proves to be best in any circumstance may be surprising: while in the above well-digging programme, using the lowest technology, a trained team was found to be preferable, in another, borehole, programme, using much more complex technology, community contributions and effort were encouraged instead: 'I believe that you can find ways of involving local people in a drilling programme. For instance, to start the drilling you need water. The drilling crew would have a lorry and a bowser, and it was usual for this to collect water from the river for the first twenty metres of drilling. We did away with the bowser, and put a small tank on site, and it was the villagers' responsibility to make sure that it stayed full of water. In some cases people had to walk several miles to keep the tank topped up. It may seem ridiculous to ask people to do that, but when foreign currency is short, and fuel costs [are] the most expensive thing on the project, there is a saving. And it is very much a part of creating a sense of ownership of the water supply.'

The idea of asking rural communities to participate in water supply programmes by making contributions during the implementation phase, whether by giving cash as in the Solomons example or in kind by providing labour, sand and gravel as in the Nepali and Indonesian examples, can also be extended to the operation and maintenance phase.

Indeed, there has been a trend in recent years towards cost recovery in water supply programmes. While capital costs paid for by the government or agency involved may largely be written off, as far as asking the recipient to contribute a fair share is concerned there has been an increasing trend towards attempting to recover recurrent costs of operation and maintenance by charging for water usage.

In many cases, particularly where water is supplied at public water points, charges are notoriously difficult to collect (for instance charging per bucketful) and the principle may have to be relaxed. In some instances, people are genuinely too poor to pay and the charge merely prevents use of the good-quality water supply. Alternatively, members of the community may not recognize the legitimacy of the charge and refuse to pay.

Increasingly then, attention has been given to gauging rural recipients' ability to pay, and subsidizing water rates. In such cases, users participate in water supply programmes by contributing a part share to recurrent costs and in this way they at least ascribe a specific value to water supply.

Other parties to participation

Beyond the communities themselves, other parties will be participating in water supply projects and programmes. Just who

Installing a hand-pump in a self-help programme, Nepal. The protective concrete apron will be laid when the well-head has been correctly positioned. VSO/Sue Eckstein

these others are will vary, as will the extent and nature of their participation, which may also be direct or indirect. But the government is almost invariably one of them, especially when projects are mounted under the umbrella of major national government water supply programmes. But even where the project has been wholly instigated, funded and run by a non-government organization, it will almost certainly have to be incorporated into the national water supply programme and meet its standards.

Government participation may spread across any or all of a number of its ministries or related public agencies. Ministries responsible for programmes in which volunteers have worked have included those responsible for public works, land use, health, housing and community development departments. In some instances, water responsibility may be divided between two or more departments, though the trend during the Water Decade has been towards the rationalization of responsibility within one department.

One volunteer, who had been working on a government programme in Malawi when such a rationalization took place, noted that it revitalized the whole approach, giving an opportunity for a strategic response to water supply needs. In this case, broad community development became integrated with the specific issue of water supplies. But although such integration or co-ordination at policy level is essential, it may not be reflected on the ground, at the practical grassroots level. In Uganda, staff working at the village level on installation found themselves having to organize meetings of locally based government employees: 'We had to get people talking to one another about the water supply: the District Treasurer, [and] the Health Inspectorate. The Water Development Department had a borehole maintenance team, who had never talked to these other people either. Building up relationships between them, before any village could be involved, took time.' But where this arises, the effort required to get people together is worth it.

Communication and participation can be further complicated by the presence of an external non-government organization, particularly when this is contributing resources and manpower to the programme. In two countries where UNICEF was supplying resources and/or expertise to government programmes, volunteers found themselves uncertain as to which was their ultimate 'employer': government or the external agency? Clarity here is important. The dilemma created by its absence exercised volunteers working on aided programmes in Nepal and Uganda, where pressure to deliver water supply targets in the short term (the main goal of one 'employer') overtook the longer term need to prepare communities to maintain supplies (the main goal of the other). As one volunteer explained: 'Donors want water supplies to be provided, but perhaps they would understand if you explained it to them sensibly, if you said, "This community is not ready to have the water supply constructed because they do not understand the whole idea of community maintenance. If we go and build the supply it will work for six months and then break down and the government will then have to spend more money fixing it, because the community is not interested." ' It is clearly in the interest of both communities and government to take this longer-term view.

Non-government organizations, whether external, or local, were often felt by volunteers to be in a better position to prepare communities for installation than to support them in maintenance. A CARE project in Indonesia began by providing gravity supplies, added sanitation and health education, extended into nutrition and agriculture, and became a community development project with water as the focal point, within four years. A community health project in northern Ghana began with the idea of supporting income-generating ventures for women. Finding that poor health was associated with low income, there was also a health education programme. From these initiatives emerged a need for improved water supplies. Thus integrated approaches can grow from various beginnings, but require some community development support.

The opportunity to compare government and NGO approaches comes from the experience of a volunteer who worked in both, in Nepal: 'Working for the government programme, (for which all material and expertise is provided by UNICEF) I felt a conflict between wanting to see immediate results, and to get as much done as possible, and at the same time helping government to get its structures right and its working practices in good shape.'

In this programme, a scheme to recruit women health educators was established in the government programme, and criteria were chosen for the selection of women to work on it: they were qualified by age, experiences like having had children, ability to speak several of the local languages and so on: 'We worked out a pragmatic basis on which we could select the best people and we found that those people worked well. But the following year the government pointed out that they were being paid at a particular Grade and they needed to have the basic qualification for staff at that Grade. Those who had not achieved that qualification were dismissed.'

While it is necessary for a national programme to observe these rules, a locally based NGO can be more flexible. The same volunteer recalls that in the NGO programme: 'The main advantage is that the project is much smaller, so it is much easier to communicate with the people who make decisions. They are concentrating on the project and spending all their time on it, whereas district officials and central government officials have so much on their plate, that it takes much longer to get decisions from them. In that small organization everybody knew what was going on.'

But a problem for small organizations is how to get the technical expertise they need without having to employ too many full-time experts. Many small NGOs have found their management resources taxed or destroyed by having to deal with programmes that are on a larger scale than previous experience has prepared them for. Here the answer seems to be some networking of skills around locally based projects. In the Kenyan example described earlier, where a diocesan development committee had established a network of communities prepared for and committed to water supply projects, a water technician was presented with a list of 135 sites which community groups proposed as water sources. Daunting and demanding though the subsequent work was, in two years many of these were actually developed, and it is this programme which, in numerical terms, delivered the highest number of water supply systems, although it lacked the pressure of a government programme to deliver quantitative results.

Additional dimensions

'If hygiene and sanitation are below a certain level, improvements in the quality and possibly even the quantity of drinking water are unlikely to improve health status.'[16]

Sanitation

Most of the water supply programmes in which VSO volunteers worked had some element of sanitation incorporated in them. Although in several it was very much a secondary element, where programmes were rooted in community health services and activities it took a more central role, and this can only be sensible if maintenance of clean water supplies is the goal. Once again, however, people's habits become the subject, and in this area, particularly, habits are difficult to change.

In Nepal, where sanitation was not originally part of the water supply programme, it has become so, and a volunteer sanitation engineer has been working on a pilot project to train women health promoters and encourage the building of latrines in villages. However, 'To make sanitation work you have got to get 70 to 80 per cent of the people in the village using latrines on a regular basis, and this is an unlikely target: there were no latrines at all when we arrived in the village.'

Teaching by example and by demonstration was one technique used by several volunteers in this situation: 'The only way with sanitation was to show people what to do. Any time we arrived anywhere, the first task was to build a latrine. We dug it very carefully. Some of them were real showpieces, and people would gather round and ask why we were doing it. In one village every single family started a latrine afterwards, but I don't know if they finished them.'

Encouraging by example was also done through the ostentatious washing of hands after using the latrine. And even blackmail sometimes entered into this. One volunteer described how he only agreed to construct an extra tapstand if the family would build a latrine in return.

But building is not the same as using, and several volunteers noted that an unhygienic latrine could be a greater threat to hygiene than former habits. In some areas, where community rules about hygiene were carefully observed and sensible, it seemed that traditional habits might well be preferable to a dirty latrine. 'Unless you get sanitation right, and back it up with health education, it is better not to do it at all,' commented an Indonesia volunteer.

In some cases, a sanitation programme preceded the building of the water supply. In a programme in Indonesia it was seen as a way of testing the commitment of a village to the

Final positioning of the hand-pump in a community-based water supply programme, Nepal. The concrete apron around the pump enables the community to keep the area clean and minimise contamination of utensils and water. VSO/Sue Eckstein

whole programme, and only when all houses had built a latrine was work on a water supply begun. Other household habits could lead to health hazards: 'All the houses were built on stilts, and the traditional way of getting rid of water from the kitchen was to throw it through the slatted floor. We tried to design a sink with a pipe down to the ground which channelled waste water into a soakaway, so that it didn't lie around as a breeding ground for insects.'

Health education
The benefits of a sanitation programme are less visible than those deriving from the actual supply of the water, yet the maintenance of water supplies could depend upon hygienic sanitation. The linking factor is health education, and in those programmes which are based on such an education programme, sanitation is often introduced first.

The basic method in all of these programmes was the identification of health workers, who were trained, and then worked either in a home village or to service several small communities, perhaps themselves then training others. Given the central role of women in sustaining family health, the recruitment of women for this task in several programmes seems logical. In a project where health education was introduced at the outset, the village committee was asked to appoint a dozen or so people to carry out health education. They would be trained and then themselves train other people in the community.

But the training of community health educators needs to be done in a way that is appropriate to the area in which they live and work. To prepare for this, a project in Indonesia conducted 'baseline' surveys in each village where requests for water supplies had been made. It would then develop a training programme based on the data collected. It was found – and this is a common finding when a research tool is used as an introduction to a community – that by asking villagers questions about health, sanitation, drainage, and so on, an opportunity was given to villagers to ask questions in their turn. 'From the relationships which built up, we were able to ask people if they would like to volunteer to be a health educator. Seventy per cent of the village volunteers we recruited in that way were women, and we encouraged them to operate in the same way themselves: to go out and survey the problems, and see where action was needed.'

In this example the health educators remained in the village during and after the installation of the water supply, and were able to reinforce earlier health lessons, with practical advice about water usage and hygiene. Working with children was a central aspect of this reinforcement, and well planned community programmes always included the installation of latrines and tapstands at the local school, backed up by classes and entertainments to encourage their proper use.

Given that, as noted earlier, community participation in the planning of the water supply is desirable, it appears to be more likely to occur and be effective if pre-supply project health education activities have been mounted. The benefits of such an approach lie, in other words, not just in improving health, but in acting as a starting point for participation. Good design of tapstands, for example, relies on some understanding of the dangers of water which has not soaked away, and is allowed to lie, attracting insects. Understanding why this design is necessary is the best basis for community maintenance of a hygienic area around the tapstand. Once the water is there, people are less inclined to listen, for the objective has apparently, in their eyes, been achieved.

All the evidence from volunteers points to the interdependence of the three elements, of water, sanitation and health education, and to their combined value in increasing popular participation.

Irrigation and agriculture
Rather separate, and only occurring in two programmes, were efforts to exploit the water supply to improve agriculture. Significantly, this occurred in integrated, community-based projects, which also had well developed health and sanitation elements in them. This description, from a volunteer in Indonesia involved in such a project, illustrates the growth of an idea: 'The irrigation started from trying to stop the stagnant water around public taps, which was occurring because of a design fault in the early ones. A soakaway was tried, and then came the thought, "If we are getting rid of water, why not try to do something useful with it?" Ideas like that were not something that came from the project manager, they came out of discussions and suggestions from everyone in the village, as we looked for good ideas. So, in some instances, instead of having the soakaway, we tried running the water to a vegetable garden. That was possible in areas where there was a good overflow. In one village where water was used in that way, the village was earning about $100 per month from the vegetable garden. But in other areas amounts were smaller, and there was not the potential to use water for irrigation.'

While this illustrates a valuable income-generating side effect of a water supply project, the primary impact from a vegetable garden is more likely to be on the nutrition levels of the people. Health education which points to these benefits is an essential once again, for otherwise another type of situation may arise: 'In some areas, where vegetables were cheap, people did not want to eat them. There were cultural reasons why cheap food was not considered desirable.'

Sustainability

Water committees

'Just as no planner of any water undertaking would allow construction to start until the adequacy and reliability of the proposed source has been thoroughly tested and proved, so should the system of management be laid down and approved by all concerned at an initial stage of planning.'[17]

The overall goal of a water supply programme is that it should not only provide good water in sufficient quantity, but continue to provide it. Where good practice has been identified throughout this account, it has been directed to this end: to ensuring that the water continues to be available in good quality and quantity, and continues to be used for the maximum benefit of the whole community. This is a collective responsibility, and those elements in planning and construction which contribute to collective working and collaboration, are liable to strengthen the sense of 'ownership' which the community feels for the water, and the sense of responsibility it has for caring for the water.

Where water committees are formed during the preliminary development work with a community, they will very likely become the bodies responsible for maintenance, or at least for ensuring that continued maintenance occurs. To do this, they will nominate the people who are to work as caretakers of the water supply, arrange for payment or maintenance of caretakers, and establish and collect contributions to a maintenance fund. They will make sure that regular inspection and monitoring of the water system takes place, with minor repairs and maintenance carried out where necessary. Major maintenance problems may need outside intervention, and it is essential that the water committees know where to go for help, and that they receive a quick response to requests.

Water committees will vary in size depending on the size of the community, but are typically four to eight people, with a chairperson, a treasurer and a secretary. Ideally there will be men and women involved, although, as has been noted, the involvement of women may occur in various ways. But their involvement is of critical importance.

Water committees may manage a maintenance fund for the water supply, with the size of contribution to this from each family agreed before construction begins. As the money is used, further funds may need to be collected, and this may either be done by levying a regular set contribution from each family, or by special collections when a repair is necessary. The latter approach seems less desirable, since repairs may often require the purchase of spare parts, and it is best if supplies of these are purchased before emergencies occur. The committee will also purchase and care for the tools required to maintain the system.

'Caretakers' appointed by water committees are responsible for carrying out minor maintenance and periodic inspection of the water system. Caretakers need to live in the village and to have been involved in the construction of the water supply. They should get basic training from the workers who have constructed the supply. Some communities may have two or more caretakers.

Another approach is to appoint caretakers for each tapstand or handpump, and an overall 'supervisor' who checks the source and pipeline, if it exists. This method has the advantage of involving larger numbers of people in maintenance and giving responsibility to those who use a particular tapstand, but it is more difficult for the water committee to monitor the whole system and to supervise the caretakers.

Caretakers are usually paid in some way for the work they do, and are held accountable for any failure to do it. In some communities, the caretaker is paid from the maintenance fund for every repair, while in others the caretaker is not required to pay the user fee charged to other villagers. In some circumstances the whole community may give days of work on the caretaker's farm in lieu of cash payment.

But whatever the precise nature of the arrangements, it is clear that the functions and duties of the water committee and the caretakers must be understood and accepted by the local community before construction takes place. In some programmes this is formalized in a written contract between village and government authority, in which the technical requirements for maintaining the system are described. This formal approach, though recommended by external agencies, is not widespread in practice, however, partly because it obviously depends on the existence of good literacy levels among the members of the village water committee.

However, the essence of these agreements, whether formal

or informal, is that the villagers understand that there will be continuing demands from the water system, and are prepared to respond to these.

Training workshops

Not all programmes offer continuing training to caretakers and water committee members, and even in small and remote communities these individuals may change. Several volunteers had established short workshops, gathering together caretakers and some water committee representatives, and giving short courses on system maintenance. Caretakers in particular need to know how to monitor the system and its use; what to look out for which could lead to damage of the system (such as animal or human behaviour); and how to counteract such threats, by, for example, regular cutting back

Discussing problems of hand-pump maintenance with women from the community, Malawi. VSO/Erica Lewis

THE PROCESS AND THE PRODUCTS 41

of undergrowth near pipelines or tapstands, clearance of soakaways, and so on. This kind of training requires resource commitment from government or other external agencies, with allowance for this made in programme budgets.

Tools and parts
The responsibility for providing tools and parts may have been accepted by the village water committee, but if they are difficult to obtain, or even unobtainable anywhere in the country, the committee may lose credibility without necessarily deserving to do so, and thus not only the actual supply system will suffer. So an important role for governments and external agencies is to make sure that parts and tools are available at all times, and at a price that poorer communities can afford.

In this, the appropriateness of the technology chosen is important. It needs not only to be appropriate, but to continue to be produced and available. Many systems in developing countries have broken down because they have been superseded by some new model, and parts are no longer available for the old. Several volunteers described how they had tried to invent ways of getting obsolete pumps and other equipment to operate, but village caretakers may neither have the tools nor the training to improvise.

There are an increasing number of industries in developing countries which are producing hardware for water systems, especially handpumps. Local manufacturing follows different options: the production of components and parts which need frequent replacement; local manufacturing of components using imported parts, like bearings; local manufacture of complete products, like handpumps. 'India, where about ten factories produce the standard India Mark II pump, and where spare parts are readily available, is a good example of successful local manufacture and maintenance.'[18]

Indeed, the success of the India Mark II handpump is notable. Its durability and relatively simple design and maintenance needs make it a cost-effective option suitable for village-level maintenance. Apart from its adoption throughout India, it has spread to thirty-five developing countries, and 100,000 are estimated to be in use, whether in UNICEF, UNDP, government or NGO water supply programmes.

Evaluation

The most useful evidence to show which methods of establishing water systems are most effective would be the systematic and comparative evaluation of different programmes. Although work has been published during the Water Decade to facilitate evaluation, and although programmes have been evaluated, there are as yet no definitive judgements on what works and what does not.[19]

Even simple evaluations can yield useful information. For example, one VSO volunteer carried out a consumer survey of villagers in the Solomon Islands to discover whether they considered they 'owned' the government-installed water supply. The response showed that people did feel this, but nevertheless continued to believe that government should repair and maintain the supply. Many supplies had fallen into disrepair as a result.

Evaluation is a tool for project management and improvement, and if the staff involved in installing supplies are also introduced to the idea of evaluation, they are more likely to use it to modify and improve the programme. Training material developed by UNICEF and IRC is available.[20]

4
Four case studies

The analysis provided in the previous chapter has been drawn from the experience of a large number of volunteers. This chapter provides four more detailed accounts of water supply projects and programmes in which volunteers have worked. The four accounts have been selected to show how diverse are the settings in which volunteers have worked, but how, nonetheless, common features (which have been analysed and discussed in the previous chapter) and lessons (set out in summary form as guidelines in Chapter 2 and conclusions in Chapter 5) emerge.

Malawi: discovering the benefits of flexible, participative approaches

Imelda Richardson had an MSc. in hydro-geology and some months' experience working on a groundwater scheme in Britain. She went as a volunteer to Malawi in 1980 to work in a government-run and -funded water programme. At the same time two experienced hydro-geologists funded by the Overseas Development Administration had joined the government team, to work with Malawian staff on special projects. The presence of a volunteer allowed a Malawian staff member to undertake extra training abroad.

During her first year, Imelda travelled around the country responding to requests from villages for boreholes, and finding suitable places to site them. She had, however, no role in the construction stage of the water supply. But during her travels she could see that there were problems with boreholes which had already been drilled. Many were breaking down, and the centrally based maintenance team was unable to cope with the demand for repairs.

The government recognized these difficulties and the programmes began to move into a second stage as a result of the lessons learned in the field, and the result being a more flexible and cost-effective programme: 'Some of the changes were technical, improving the design of boreholes, and using local materials. But also, the methods were changing. Instead of using the hydro-geologists just to help determine the site of a borehole only, we started to be involved in the supervision of the actual construction, showing drillers when the hole was deep enough, doing basic tests as drilling was going on, so that you could identify where the main veins were.'

Having a hydro-geologist advising drilling teams also meant that early judgements could also be made about whether drilling was the best solution to water need, or whether another way of providing supplies might be more suitable, and cheaper. So from a programme which was based on the idea of drilling so many boreholes per head of the population, there emerged a different approach: projects serving areas of up to 50 000 population co-ordinated by a geologist or hydro-geologist, who would work with local communities to discover their water needs and preferences, and who would then supervise contractors and village labour to construct the system.

In her second year Imelda managed a pilot project to explore this new, more flexible, way of working, in which there would be more consultation with local people, and participation by them. The new approach also involved a demonstration process, covering an area where twenty-five water points were planned, each serving 250 people. Imelda's initial task was to set about discovering what people wanted, and what was technically possible. 'I started in a few villages, talking to people, and then, once construction started, I would move on to begin the negotiating process in the next village: a staggered programme. It was all experimental, finding out how much people were prepared to contribute. The actual digging and drilling were going to be carried out by a team, but there were things that the villages could do: collecting sand to make concrete rings for lining wells, feeding the workforce, helping in the construction of surrounds.' But looking back on this part of the work, Imelda felt subsequently that she should have asked village leaders to create and run water committees. 'At that stage I didn't know as much as I know now.'

Nonetheless, it was possible to mobilize community support, and Imelda learned that the extent of community support and participation depended largely on the strength of local leadership and the respect in which it was held.

Initially, women were not adequately represented in this

process, and Imelda feels that water committees would have offered a way of guaranteeing their involvement. So this, too, was subsequently incorporated into the programme. But other factors also came to affect the extent of participation of women in the programme: 'I know I should say that I was able to get women in the villages involved in water supplies because I am a woman myself, but I don't think that was the case. What did make a difference was that I eventually got a Malawian counterpart, a geologist. And the department was fortunate enough to recruit a woman to the post. I think having a Malawian woman doing the job was very useful – a very powerful symbol indeed.'

As time went on (after Imelda's departure) the pilot project was expanded to cover an area with a population of 200 000. By this time it had become clear that, in a country with poor roads and a lack of road transport, central maintenance was not working. Maintenance had to be community-based. Later modifications of the handpump which had been used on the pilot project meant that it could be dismantled with one spanner. 'I didn't have any part in that, but it was part of the process that was starting when I was there. The woman geologist who took over from me became the leader of training in the maintenance of handpumps, and trained local committees of women to implement the maintenance programme. That proved a successful formula: the fact that women were being trained in technology by a woman overcame their reluctance to get involved in something that was seen as a man's domain.'

Ten years on, the latest news of the first source constructed under the pilot project is that it remains in operation. The handpump that was originally installed has been changed for a modified, even more easy-to-repair version.

Indonesia: building in the health component, and then building on it

Robert Coe had a diploma in public health and eight years experience working for local authorities in Britain when he went to work as a volunteer in a water supply programme funded by CARE, an external NGO working in several areas of Indonesia. Before his arrival the project had concentrated largely just on water supply. Robert's role was to develop the public health aspects of the programme, by training local field staff.

To begin with, Robert assessed what was happening in villages, and then devised a programme that related to the practicalities of water installation. In particular, it meant training field staff in their health education roles.

'In each area there was a manager and an assistant and eventually about ten field officers plus a surveyor or draughtsman. The field officers were based almost full-time in the villages in which water supplies were being constructed, and then they would move on to the next village. They saw through the installation and worked on sanitation and health education. Training them also meant working together with them to devise programmes that they could use, including health education materials and so on.'

The approach also meant that each village would appoint or select a 'cadre' of about twelve volunteers to carry out health education work and continue it once the supply was built. The field staff were also responsible for training these groups.

Typically, entry into a village was marked by a consciousness-raising exercise. 'We had the use of a vehicle and our own projector and generator for showing films. A whole village would turn out on a Saturday night to watch the film. I don't think that was much use in training people, but it did arouse an initial interest and gave field staff an opportunity to raise questions about water usage.'

A basic survey of each new community was then conducted by the field staff, giving some idea of levels of health and of the facilities available. It was felt important to ascertain this in every area, and to spend time understanding each community, because there were thirty different local languages and many differences between ethnic groups. Robert acted as an adviser to the field staff in this process.

He also devised materials for the field officers and their volunteer 'cadres', drawing upon a respected source, *Helping Health Workers Learn*[21] in devising materials: 'It is packed full of ideas and suggestions from all over the world; we tried them out to see what worked. And everybody who was working on this tried out things they thought might be effective, including the volunteers in the villages. All the time we were trying to get away from the rather formal learning patterns that people had been through in school.'

The village health volunteers became responsible for monitoring health in the villages, carrying out monthly weighing of young children in order to spot cases of malnutrition. The project supplied weighing charts, and every village had a set of scales for weighing rice before it was taken to market. But the volunteers needed continuing support, after the construction of the supplies was completed, since it was easy to lose enthusiasm at this stage. 'The villages where interest flagged were places you could almost predict.'

Signs that a village would remain motivated and enthusiastic about health and sanitation, even after the supply had been installed were: experience of previous collective action in the village, a respected community leader who could get people organized, and a high turn-out at preliminary meetings. The field staff, Robert felt, acquired an almost intuitive sense of where interest would not just be maintained but be capable of being built upon.

This was important in the progressive expansion of the project, from basic water supplies, via the health education work, to broader community development in a number of areas. 'In three years the programme moved from water supplies and nothing else to a water supply, sanitation, health education, income-generation and agriculture programme: a whole range of things, all based around the water supply.'

Some of those developments were directly linked to water, but others were a product of the community mobilization that had occurred. 'The programme showed that going into a village and providing water supplies alone was not the end of the process. To do that was fine, but why stop there? If your aim is to improve people's standards of living and quality of life, why go part of the way, and not further? The programme was motivating people to work together, so why not build on that to help them work together in other ways.' Other ways included gardening projects, garlic production and the introduction of bird-rearing for eggs.

Robert emphasizes that this kind of community development was possible because the water supplies were an attractive entry point for the work. 'If you go into a village and say, "Provide materials and we will show you how to build latrines," nobody will be interested. If you go in and say, "We will provide technical knowledge if you supply materials and labour to build a water system," people will be eager to start.'

Ghana: from community development to water supply

Previous experience as a community worker in Scotland working with mothers of pre-school children proved a useful background for a VSO volunteer working with the Binaba Community Health Project in northern Ghana. But Fran Loots's job description was a vague one: the Royal Commonwealth Society for the Blind, which was supporting work in the area, felt that work on health education and the condition of women, would have an impact on blindness. But the rest was up to Fran.

For the first six months Fran explored possibilities, reporting to the local management committee of the Binaba Area Community Health Project. She carried out a health education programme in primary schools, developing games and teaching aids from local materials, getting children to relate what they had learned to one another, and putting the emphasis on practical action in daily life: covering faeces, washing hands, washing eyes daily and so on.

At the same time she was meeting with groups of women who had been referred to her by the local health superintendent. 'To start with they were not talking about water. They were keen to have an income-generating project, and wanted to grow more food, so that they could sell it. In fact, it was clear that a lot of the women did not have enough to eat for themselves.'

Water emerged as a source of concern: 'The borehole supply does not meet the needs of the area and traditional sources dry up in January/February,' she wrote in 1987. 'I hope to combine education about water with practical steps to improve the situation . . . I feel that the interests of the men and women are different when it comes to water. The men want a dam for their animals, the women need good water for drinking and washing.'

In order to hold women's meetings without alienating the men in the community, Fran visited the elders in the villages and explained why they were necessary, because she did not want to interfere with existing cultural patterns. As water committees came to be established in the villages they were composed of three men and three women, for 'men were needed, too'.

In some villages, committees took six months to establish, while in others it was much quicker. 'It often depended upon a couple of people who knew how to organize others. If the process was taking time, they were simply left to get on with it. We would just say, "We'll come back when you are organized." ' Once word got around that water improvements were possible, the project was inundated with requests. Priority was given to villages with no source of water, or where the young people had left and the population was frail and ageing.

But what form would the supply system take? It emerged from local knowledge and experience, expressed through the committees, that some thirty to forty years previously there had been many wells dug in the area and lined with laterite stones. 'Older people still remembered the techniques. There were some drawbacks in the quality of the water, but amounts were good, and it seemed a good compromise and an affordable method, which we could get on with immediately.'

The first well was entirely hand-dug with no tools other than a pick-axe. Between ten and fifteen people dug the well in ten days: men digging, women feeding them and carrying sand and stones. A well-head with a circular platform was designed to slope to a drinking trough for animals. A Wateraid engineer based in Ghana advised on lining the well with concrete rings.

This simple technology was backed up by a health programme, 'usually conducted under the nearest tree'. Here, the need to keep the water clean was stressed. The children were told not to throw sticks into the well, and demonstrations were organized to show people how water could become contaminated.

There was much discussion about the kind of buckets to be used, and how the communal bucket could be maintained. Local materials were used to make buckets, and rope, though these did not always work. Locally made rope rotted and contaminated the water. 'Solutions were not easy. But the idea was that in each village people would work out the solution which suited them.'

There was a financial contribution from each household to the well fund. The village water committee collected this. 'It made people feel that the well was theirs.'

A sanitation programme grew alongside the water provision, with latrines built at schools first of all. A national radio campaign had effectively caught the imagination of the children, and pupils and staff were jointly responsible for building and maintenance, with the project providing advice and materials. Requests for latrines came from compounds in villages, too. If the family did the digging, and provided half the cost of the cement, the project would construct it, and conduct a health programme with the compound, a programme which covered how to use the latrine and how to care for it, and the health hazards that might result from a badly maintained latrine.

The project continued with another volunteer, but will eventually be run by local staff. Fran feels she learned that skills and knowledge are there in local communities, and that 'the techniques are there. What villagers need is support, encouragement and co-ordination.'

Nepal: the demonstration effect

Since 1979, VSO has been recruiting volunteers for the Nepal government's Community Water Supply and Sanitation (CWSS) programme. The programme is supported by UNICEF. Tom Mosquera, a graduate engineer, was sent to a field office responsible for three administrative districts in the far west of the country. There he managed water supply construction in two villages, serving 5000 people.

The systems constructed were of the gravity-flow type. An intake structure was built at a spring higher than the village, and piped to the village through a buried pipeline to reservoir tanks, and then distributed to public tapstands. If the drop from source to reservoir was very steep, break-pressure tanks were situated at every 60-metre drop in height. Although the construction task was straightforward in principle, as Tom explained, these projects are 'easy to build badly and can be extremely problem-prone'.

As project manager, Tom found himself co-ordinating planning, resources, village input and construction. In one village, many surveys had already been carried out, but none had been approved for funding. 'The villagers' reaction to the project was to ask how much money they were going to earn from it, as they had heard of numerous projects which, once built, did not work at all. As our project required both voluntary and paid labour, they did not want to work for nothing on a project which would yield no benefits to them. It was a difficult and constant problem to convince villagers of the integrity of the CWSS programme.' Once approval for the planned system had been given, resources were released and materials purchased. In one village, Babla, construction had begun when supplies of pipe ran out because of contract problems between the field office and the main supplier. 'Consequently pipeline work came to a halt. Construction of the intake slowed down. Eventually materials were brought by emergency airlift, but there was insufficient to complete the pipe before the monsoon. The morale of the people was low and they did not believe the project could be completed. After the monsoon much of the trench digging work was wasted as the sides of the trenches had all fallen in on themselves.'

Because of the distances over which materials have to be carried in Nepal, local people are paid for some contributions to the construction. There is a fee for portering, but this varies from district to district. The route to a second village, Kanda, in which a system was to be constructed passed by Babla, where the rates paid to porters had for some reason been higher. 'It was not long before the villagers in Kanda realized that they were being paid less than the villagers in Babla, for walking twice as far. The goodwill of the Kanda people was maintained, however, and they continued to transport enough materials to continue with the construction. I managed to secure an increased wage rate for them.'

The Kanda water supply was completed, and the quality of construction was high. 'This was mainly due to the hard work and quality control of the Nepali chief technician, with whom I worked closely. His brilliant communication skills enabled me to explain the technology to the villagers. I learned from him about local situations and the attitudes of the people.'

In the second construction season, pipe at last arrived for the Babla water system. By this time villagers were aware that water supplies had been built near by. 'I think the fact that three other systems were operational was an influencing factor in getting people to work. It was part of my thinking that we should supply a makeshift water system for special functions in a village, like weddings, that occurred before systems were completed. I felt that the goodwill created increased people's motivation and willingness to work on the main construction. Whether it did or not is debatable, but work progressed much faster in the latter half of the construction season.'

Apart from the construction projects, Tom organized training for the programme's twenty-five technicians based at the field office. 'Much of the training was refresher both on theory and practice, which is useful as it is good for quality assurance to recap on how things should be done properly.'

Another responsibility was for the maintenance of ten projects in the whole district. This task presented a problem: 'It was seen as a lower priority than new projects, because these give the impression of good capital development to aid agencies.' However, many of the more recent projects in the area had been well constructed and there has been a gradual decrease in demands on the maintenance budget.

5
Conclusions

Two principal conclusions can be drawn from the experience which has been described and analysed in the previous two chapters.

First, the main elements which will comprise, in theory at least, an 'ideal' water supply project can be identified.

Within the general context of the project, and during the preparatory stages for it, these elements will include:

- a sensitivity to, and thorough understanding of the culture and structure of the community which is to benefit from the water supply. In particular this will involve identifying leadership and 'government' structures in the community;
- securing the understanding of, and engaging the involvement of the community, its leaders, and particular groups within it (most importantly, women) in the water supply project is desirable, from the very earliest stage possible. If this involvement has stemmed from the fact that the community has itself expressed a need for new or improved water supply, and what form it might best take, so much the better;
- efforts to consult with, and raise levels of awareness and understanding in the community can usefully be built around activities geared towards health education and promotion, or alternatively, integrated with existing community development efforts and structures;
- the establishment of a water committee linked where necessary and appropriate to existing 'government' structures in the community, and to separate women's committees, will provide a clear and central focus for the project;
- the more local knowledge and skills can be utilized in the design of the supply, the better.

During the actual construction of the water supply system, the elements will include:

- wherever possible, the use of simple and appropriate

technology, which is reliable, easily maintained and repaired, in the 'hardware' of the supply;
- the involvement of the community's efforts, time, and resources in the actual construction;
- continued efforts to improve community awareness and levels of health education.

Both during and after the construction, the elements will include:

- the establishment of arrangements to ensure continued community contributions to maintaining the system, including the identification and appointment of 'caretakers';
- building around the system, and/or the leadership/committee arrangements established to construct it, other community development efforts to improve the quality of local life;
- arrangements to ensure the availability of spare parts and training for 'caretakers' and others;
- evaluation arrangements.

The second general conclusion to be drawn is that such is the diversity of cultural, social and economic circumstances in which water supply projects will be mounted, that the elements identified here should be seen as a set of guidelines to be modified and expanded as local circumstances dictate. These guidelines, summarized above and discussed in more detail earlier, appear, in the light of VSO's experience to be the ones which should be uppermost in the minds and plans of the people and agencies involved in water supply projects in rural communities.

References

1. World Health Organisation (1981), *A Way to Health: Drinking Water and Sanitation 1981–1990* (WHO)
2. Agarwal, A. *et al.* (1981), *Water, Sanitation, Health – For All?* (Earthscan)
3. Pickford, J. (1989), in Kerr, C. (ed.), *Community Water Development* (Intermediate Technology Publications)
4. Pacey, A. (1977), *Water for the Thousand Millions* (Pergamon)
5. *Ross Institute of Tropical Hygiene Report* (1988, Ross Institute)
6. *International Water and Sanitation Centre (IRC) Report* (1981, IRC)
7. Kerr, C. (ed.) (1989) *Community Water Development* (Intermediate Technology Publications)
8. For example, by Feachem, R.G. (1988) in *Rural Water and Sanitation, community participation in appropriate water supply and sanitation technologies: the mythology of the Decade* (Proceedings of the Royal Society)
9. Proceedings of the Symposium on Social and Non-Economic Factors in Water Resources Development (1976, New York)
10. Pickford, J. in *Waterlines*, Vol. 1, No. 2, October 1982
11. *Women, Water Supply and Sanitation* (1984, National Training Seminar, Addis Ababa)
12. Hannam-Anderson, C., 'Women and Water' in *Waterlines*, Vol. 2, No. 1, July 1983
13. Burton, I. (1979), *Policy Direction for Rural Water Supply in Developing Countries: Programme Evaluation Discussion Paper No. 4* (USAID)
14. OECD Development Assistance Committee (1977, OECD)
15. World Bank (1978), *World Development Report* (World Bank)
16. World Health Organisation (1981), *A Way to Health: Drinking Water and Sanitation, 1981–1990* (WHO)
17. Wood, W.E. in *Waterlines*, Vol. 2, No. 2, October 1983
18. IRC Newsletter 181 (March 1989, IRC)
19. For example, see Cairncross, S. *et al.* (1980), *Evaluation for Village Water Supply Planning* (John Wiley and Sons)

and also: World Health Organisation (1983), *Minimum Evaluation Procedures (MEP) for Water and Sanitation Projects* (WHO)
20. IRC (1988), *Evaluating Water Supply and Sanitation Projects* (IRC)
21. Werner, D. and Bower, B. (1982), *Helping Health Workers Learn* (Hesperian Foundation)
and also: Donnelly Roark, P. (1987), *New Participatory Frameworks for the Design and Management of Sustainable Water Supply and Sanitation Projects* (UNDP/PROWESS/WASH)
Glennie, C. (1982), *A Model for the Development of a Self-Help Water Supply Programme: World Bank Technology Paper 2* (World Bank)